SCIENCE FRONTIERS

MICRO MACHINES

ULTRA-SMALL WORLD OF NANOTECHNOLOGY

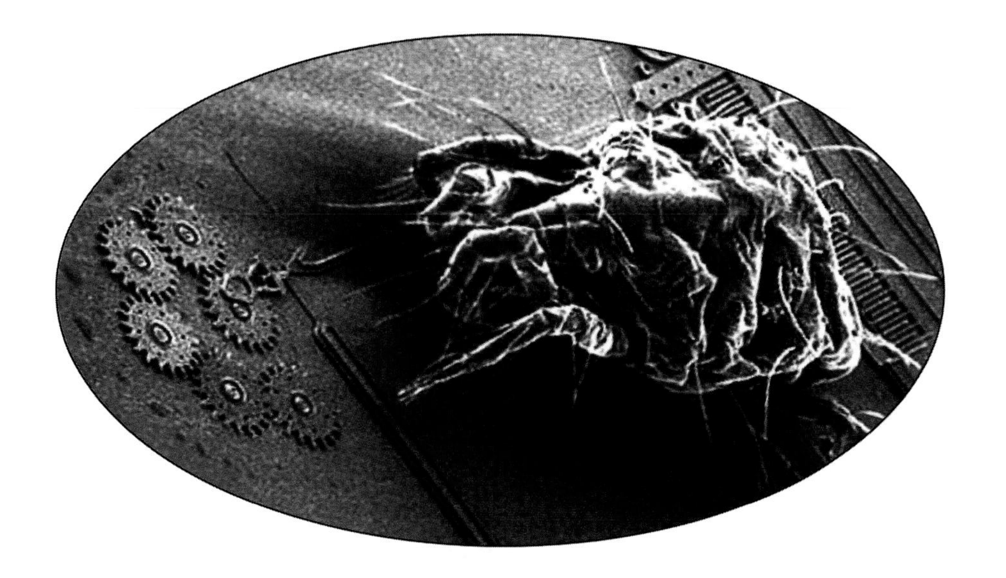

DAVID JEFFERIS

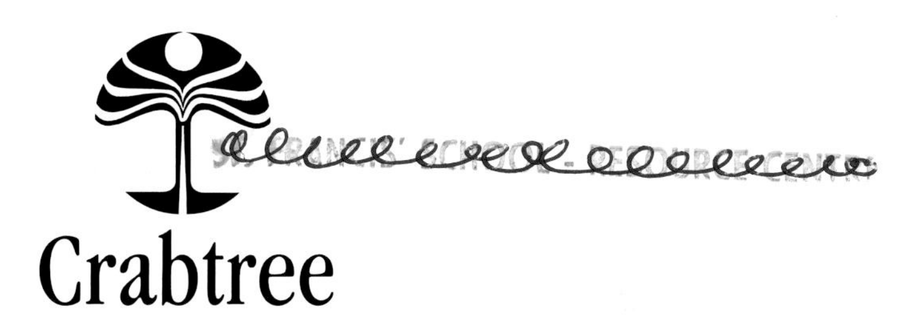

Crabtree

INTRODUCTION

This book is about things that are very small. The ability to create extremely small machines and products is the key to the new field of nanotechnology. Nanotechnology is an advanced type of **engineering** that deals with structures that are mostly too small to be seen by human eyes.

The field of nanotechnology got its name because it deals with things that are measured in nanometers (nm). A nanometer is incredibly small. One foot (0.3 meters) has over 304 million nanometers. The size of the period at the end of each sentence in this book is about 300,000 nanometers across.

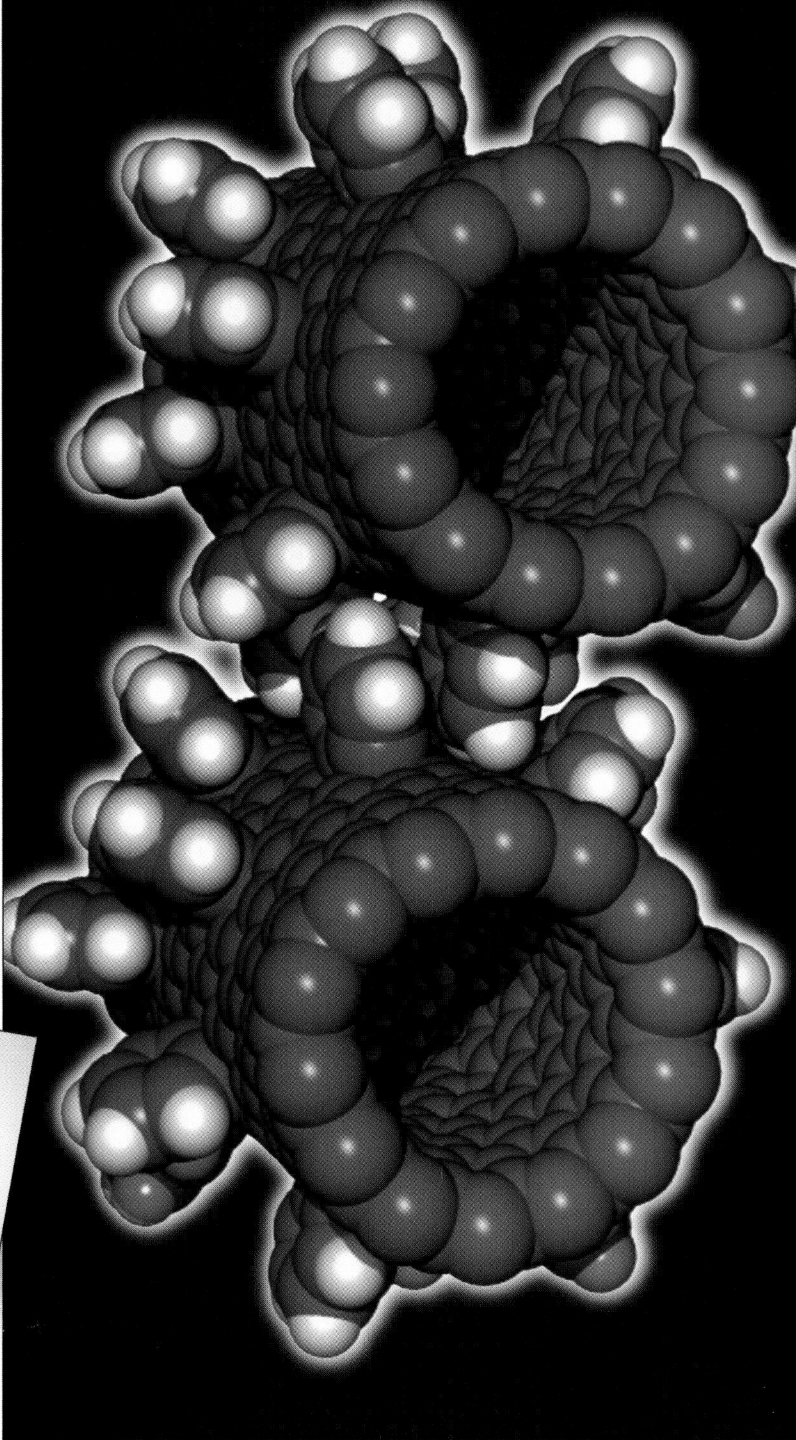

Crabtree Publishing Company
www.crabtreebooks.com

PMB 16A
350 Fifth Ave.
Ste. 3308
New York
NY 10118

616 Welland Ave
St. Catharines, ON
Canada
L2M 5V6

Edited by
Isabella McIntyre

Coordinating editor
Ellen Rodger

Project editors
L. Michelle Nielsen
Rachel Eagen

Production Coordinator
Rosie Gowsell

Educational advisor
Julie Stapleton

Technical consultant
Mat Irvine FBIS

Created and produced by
David Jefferis/BuzzBooks

Cataloging-in-Publication Data
Jefferis, David.
Micro machines : ultra-small world of nanotechnology science frontiers / David Jefferis.
p. cm. -- (Science frontiers)
Includes bibliographical references and index.
ISBN-13: 978-0-7787-2859-7 (rlb)
ISBN-10: 0-7787-2859-5 (rlb)
ISBN-13: 978-0-7787-2873-3 (pbk)
ISBN-10: 0-7787-2873-0 (pbk)
1. Nanotechnology. I. Title.
T174.7.J44 2006
620'.5--dc22
2005036726
LC

Pictures on these pages, clockwise from above left:

1 A researcher holds a model of a hexagonal, or six-sided, molecule.
2 A pair of tiny gears, each made of atoms linked into a circular shape.
3 A medical delivery system brings lifesaving drugs to a diseased cell.
4 A design for a robotic micro machine.
5 A nanotube designed to hold fullerene spheres.
Previous page shows:
A household dust mite crawling toward a set of microscopic gears.

CONTENTS

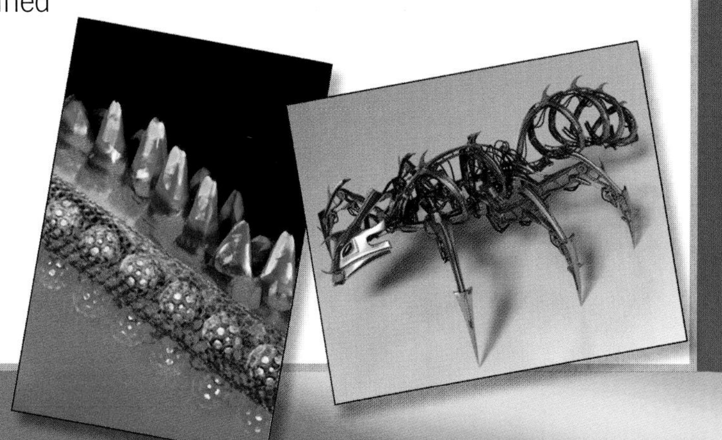

WHAT ARE MICRO MACHINES?

Computers were once as large as a room. As scientists found ways to make electronic parts smaller, computers became popular and changed the way we live. Now, nanotechnology makes things even smaller for use in micro machines.

▲ **Nanotechnology started with a talk in 1959 by U.S.** physicist **Richard Feynman (1918-1988). In his talk, Feynman discussed the possibilities of making things smaller.**

Nanotechnology works with tiny pieces of matter. Matter is any kind of substance that has **mass** and takes up space. By taking tiny bits of matter, called atoms, and arranging them in certain ways, engineers can build materials that are smaller, lighter, and stronger than any other known materials.

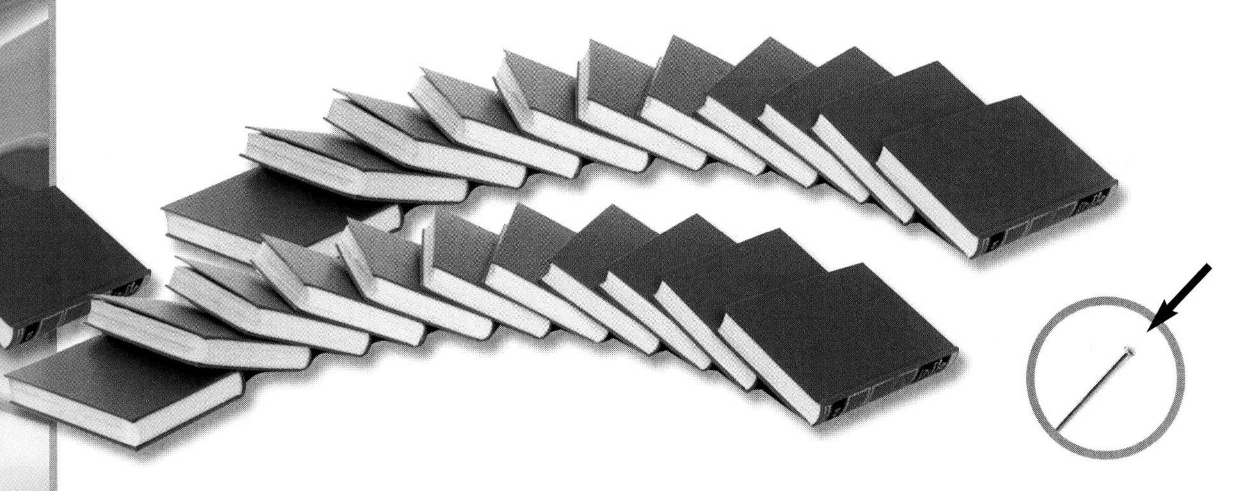

▲ **Feynman's suggestions included shrinking all the information contained in a set of encyclopedias and placing it onto the head of a pin.**

Nanotechnology can be used in a variety of ways. It has already been used to create a special coating for cloth that is then made into wrinkle-free clothing. In the future, powerful computers the size of a fingertip will be made using nanotechnology, as well as other tiny machines, called micro machines. Some people think that nanotechnology will dramatically change the way we live in the future, just as computers have done in the last few decades.

ATOMS AND MOLECULES

Atoms are the tiny building blocks of matter. Everything in the universe is made of atoms. Elements such as hydrogen and oxygen are made of just one kind of atom.

Atoms also cluster together in groups of two to thousands to form molecules. For example, a molecule of water is made of two atoms of hydrogen and one of oxygen.

FACTS AND FIGURES

> Most red blood cells are about 7,000 nanometers wide.

> A person's fingernails grow at a rate of about 1 nanometer per second.

> Atoms are usually about 0.1 to 0.2 nanometers wide.

> New materials made from nanotechnology consist of no more than 400 atoms.

▽ There are one billion nanometers in three feet (one meter). To understand just how small a nanometer is, imagine that the marble is one nanometer. It would take one billion marbles to cross the Earth's diameter, or width.

The Earth is 7,926 miles (12,756 kilometers) across

A marble is 0.5 inches (12.8 millimeters) across

◀ Human hairs vary from fine blonde strands that are 17,000 nanometers wide to firm, black hairs that are about 180,000 nanometers wide. A typical strand of hair is about 80,000 nm wide. Most nanotech engineering is carried out at sizes up to about 400 nm, about 200 times less than the width of the hairs shown in the highly magnified big picture at left.

BUILDING SMALL

How do scientists make things that are too small to see? Here is a look at some of the methods used.

Most companies that use **nanoparticles** to make products keep their expensive research efforts top secret! The nanoparticles they use have to be created before they can be added to exisiting materials. By adding nanoparticles, some materials are made stronger.

▶ **Nanotechnology is an area of science that is expanding quickly.**

▲ **K. Eric Drexler (1955-) is an American engineer who popularized nanotechnology in his 1986 book, *Engines of Creation: The Coming Era of Nanotechnology.***

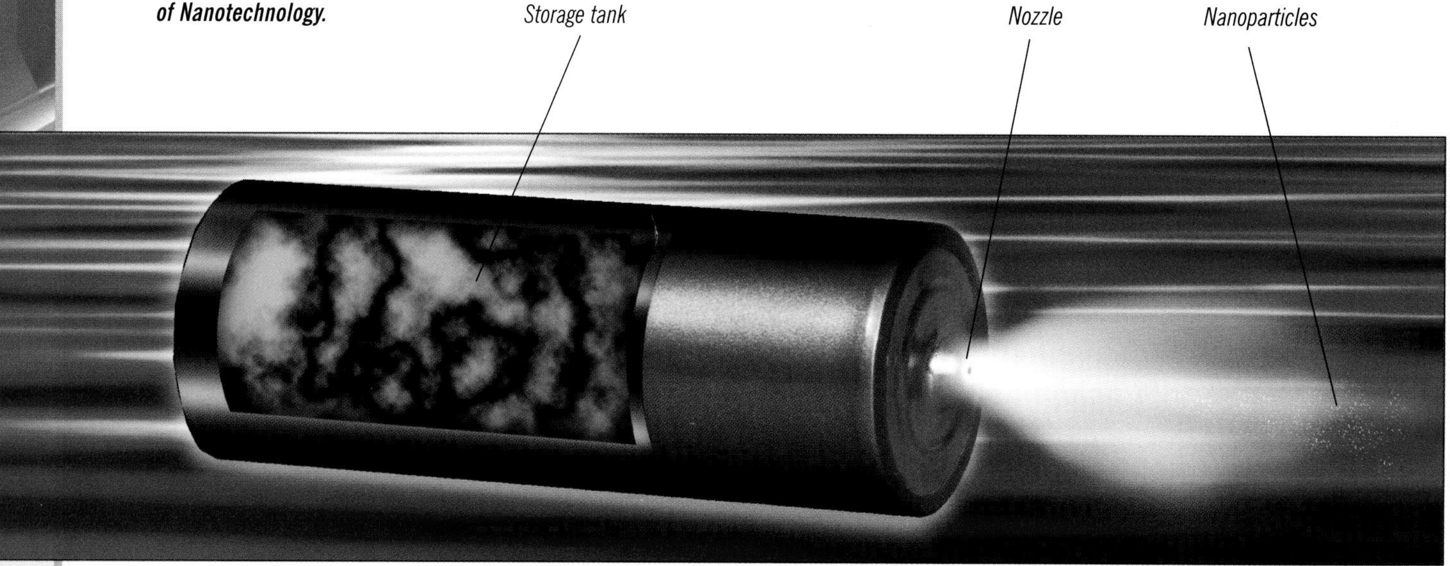

One method of making nanoparticles uses **carbon dioxide** stored under heat and pressure in a special tank. A material, such as **teflon**, is dissolved in the tank. The dissolved material and carbon dioxide are sprayed through a nozzle at one end of the tank. The carbon dioxide evaporates instantly as a gas, and the material squirts out as a spray of nanoparticles that are collected, ready for use. Many materials can be used to make different kinds of nanoparticles. Teflon nanoparticles are used to coat medical equipment used in heart operations.

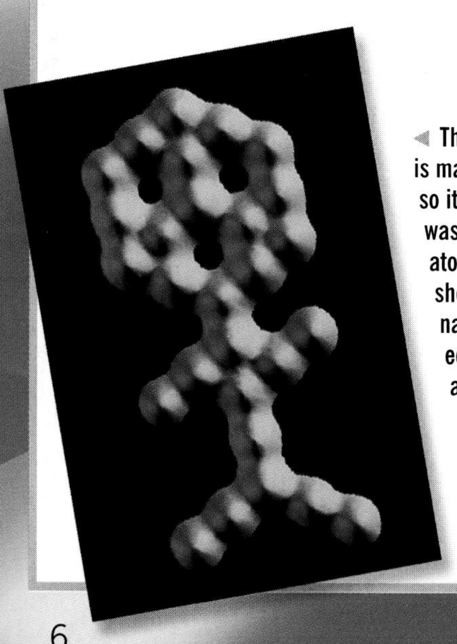

◀ **This little stick figure is magnified many times so it can be seen here. It was built using just 28 atoms and was made to show how precise nanotechnology equipment is at arranging atoms.**

NANOTECH'S FINEST INSTRUMENT

The atomic force microscope (AFM) is nanotechnology's basic tool for probing, or examining, surfaces.

The AFM works like an old-fashioned record player, in which a needle tracks along a groove, replaying changes in the groove's shape as sound. The AFM does much the same with surfaces.

The difference is that the AFM has a tip small enough to explore the nanoworld. The AFM can also make nano materials. Molecules are dropped on a surface using the AFM's tip.

Another way to make nano objects is by etching, or cutting out, molecules from a surface. To apply molecules to a material, a special stamp can be used.

◁ The tip of an atomic force microscope.

MICROSCOPIC MEMS

▲ This robot came before MEMS. It was small enough to crawl in a body's digestive system, where it took pictures using a tiny movie camera.

Micro-electrical-mechanical systems (MEMS) **are miniature machines. They are etched out of small pieces of** silicon **called wafers and have moving parts.**

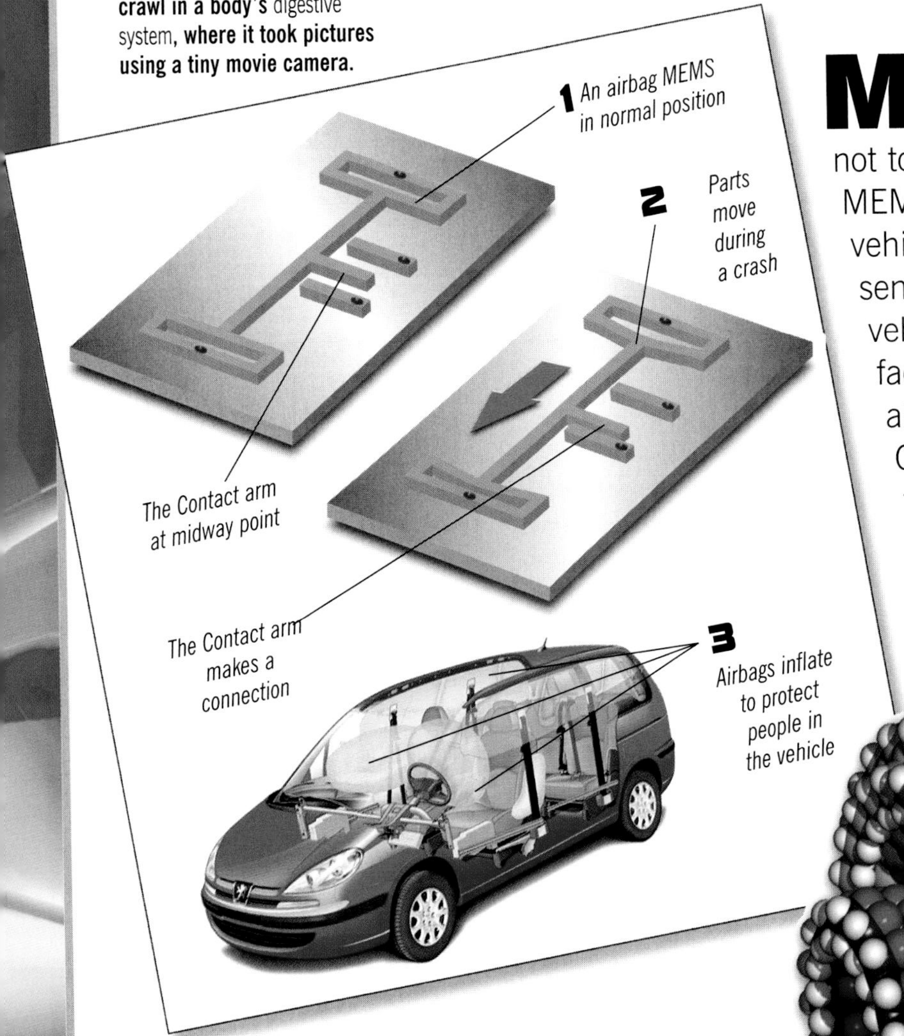

1 An airbag MEMS in normal position

2 Parts move during a crash

The Contact arm at midway point

The Contact arm makes a connection

3 Airbags inflate to protect people in the vehicle

MEMS are not considered nano-sized because they are just large enough not to be invisible. One of the earliest MEMS was the **sensor** that sets off vehicle airbags. The sensor is extremely sensitive to the energy created when a vehicle brakes, and very reliable. These factors made it an instant hit. Today, almost all cars use this MEM design. Other MEMS are also used today, such those found in camera electronics. There is even a MEMS used to detect smells.

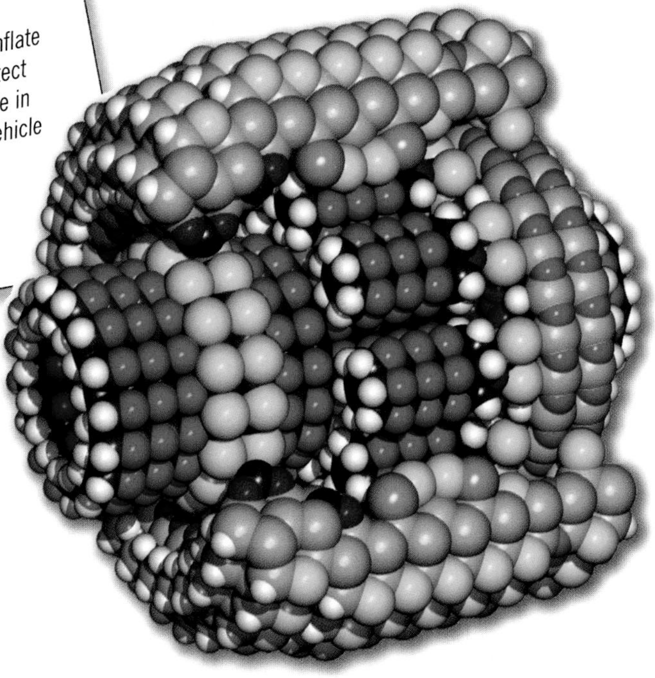

▲ The MEMS airbag sensor is very simple. A sudden braking movement jerks it out of shape. Two contact pieces then touch, creating an electrical signal that tells the airbag to inflate.

► Engineers are working on making nano-sized MEMS that use molecules as mechanical parts. This design is for a ball bearing, a common part in many machines, just five nanometers wide.

Another plan for future MEMS use is to help save the amount of energy used in a building. Hundreds of MEM sensors would constantly check temperatures. Computers would then be able to adjust the temperature in only those areas that needed it.

A set of gears in a MEM system is shown in this microscope photo. The gears are made out of a sliver of silicon.

This picture compares the size of a dust mite to MEM gears. The mite itself is very small, about one-third the width of a human hair.

WEIRD WORLD OF NANO

It is hard for most humans to imagine how things work in such a small scale as nanotechnology. For example, heavy objects often sink in water. In the smaller-sized world of insects, some insects can easily walk across the surface of water.

Nano-sized substances differ in surprising ways from their larger-scale forms. A liquid like water can seem like a thick, gooey substance to an insect.

Differences due to scale are also true for nanoparticles. Iron nanoparticles are stronger and bend easier than the full-size material. One nanotechnology research team found that a group of gold molecules can move like an amoeba. In nano-form, materials are often stronger than their bigger forms because they are too small to crack or split.

An insect walks on water safely, supported by surface tension.

COPYING NATURE

Biological systems **are nature's own micro machines. Living organisms are built from tiny** cells **that are always changing, or evolving. Engineers gain useful ideas from studying nature when designing new nano systems.**

By studying **bacteria**, researchers have learned how to build nano machines that can move around. Many bacteria are propelled by a tiny biological motor that spins their tails, known as flagella. As flagella whip around, bacteria move quickly through a liquid, such as blood.

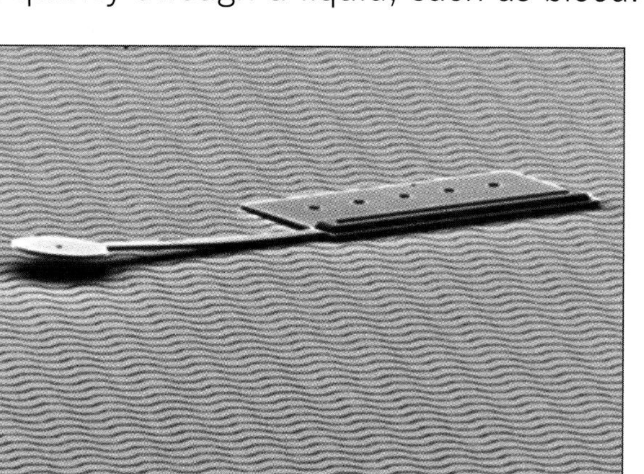

The microworm nano system has a rudder **that helps with steering, allowing it to move to the side and make turns.**

A Gecko's foot is covered with 500,000 tiny hairs called setae

Another idea for nano system movement is taken from the inchworm. The microworm is made from a tiny slice of silicon, and is steered by a rudder that moves it around corners and spins it in circles. The microworm can even push specks of dust out of its way. The team that built it thinks that a future version will be good at cleaning inside electronic circuits.

The tips of the setae split into even tinier hairs, called spatulae

The secrets of the gecko's sticky feet were a mystery until researchers solved the puzzle in 2002.

Nano system researchers use nature in other ways too. One way was to try and recreate the sticky hairs on the feet of geckos to make a super sticky tape. A gecko can hang upside down by just one foot using the millions of tiny hairs, each no more than 200 nanometers wide, that cover the bottoms of their feet. They stick due to an attraction between molecules called the **Van der Waals force**.

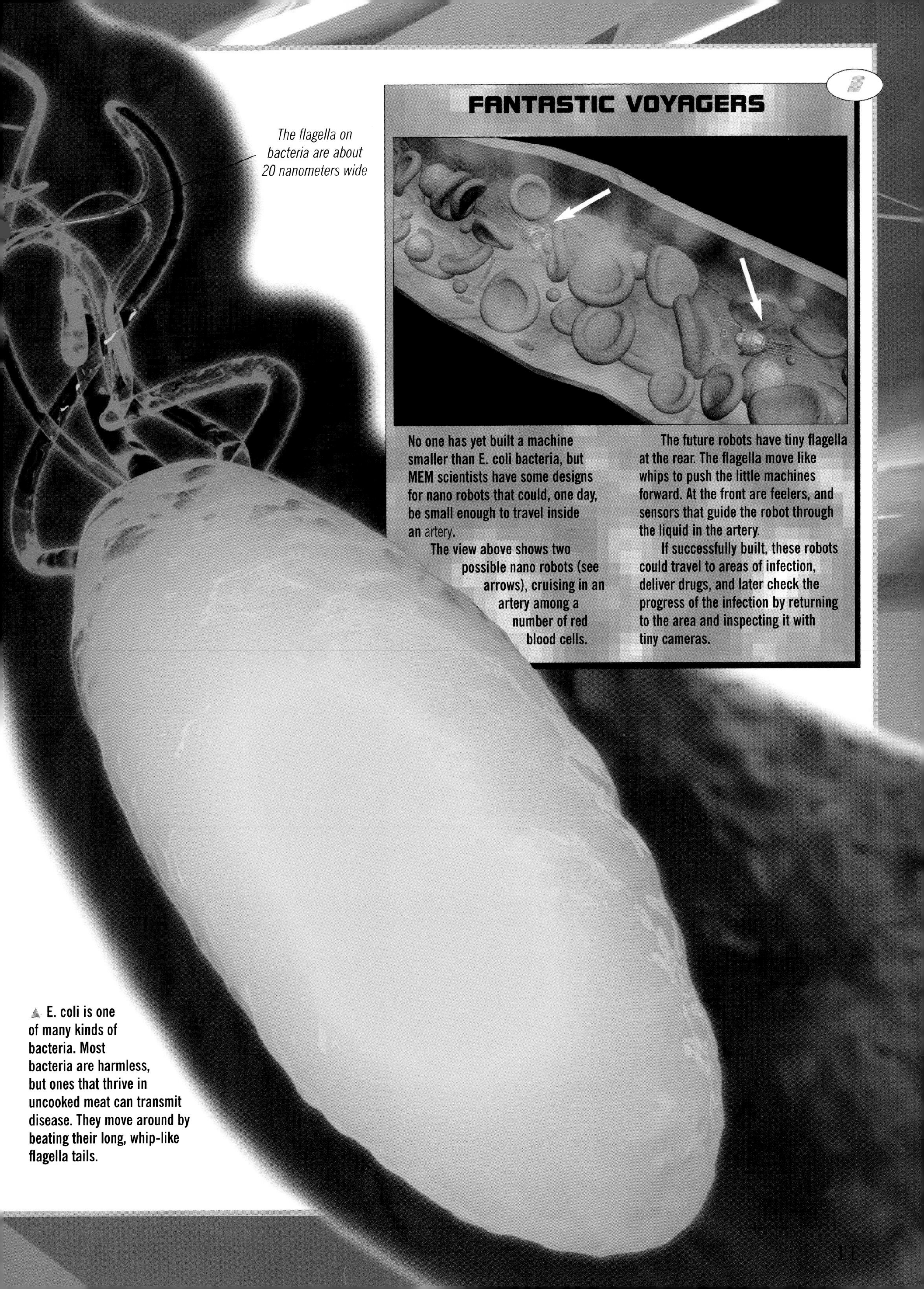
The flagella on
bacteria are about
20 nanometers wide

FANTASTIC VOYAGERS

No one has yet built a machine smaller than E. coli bacteria, but MEM scientists have some designs for nano robots that could, one day, be small enough to travel inside an artery.

The view above shows two possible nano robots (see arrows), cruising in an artery among a number of red blood cells.

The future robots have tiny flagella at the rear. The flagella move like whips to push the little machines forward. At the front are feelers, and sensors that guide the robot through the liquid in the artery.

If successfully built, these robots could travel to areas of infection, deliver drugs, and later check the progress of the infection by returning to the area and inspecting it with tiny cameras.

▲ E. coli is one of many kinds of bacteria. Most bacteria are harmless, but ones that thrive in uncooked meat can transmit disease. They move around by beating their long, whip-like flagella tails.

FANTASTIC FULLERENES

▲ The best-known invention of Richard Buckminster Fuller (1895-1983) was the geodesic dome, in which metal tubes are linked to make a lightweight but very strong structure.

Fullerenes are strong, hollow spheres and tubes made of carbon **atoms. They are named after American engineer Buckminster Fuller, who built large domed structures that were shaped similar to fullerenes.**

Carbon atoms exist in various forms in different substances such as in graphite, or the soft, black material used in pencil leads, and in diamonds. Fullerenes are another form of carbon molecules that have recently been discovered. In fullerenes, the carbon molecules are joined end-to-end to make an incredibly strong and light structure. Fullerene spheres, also called "buckyballs," are just a few nanometers wide, about 50,000 times less than the width of a human hair. Buckyballs can be opened up, rolled out like chicken wire, and formed into tubes, called carbon nanotubes (CNTs), that are ten nanometers wide.

► Carbon nanotubes are many times stronger than steel, but weigh six times less.

Fullerene spheres are used for their electricity-conducting properties, which are far better than traditional copper wire. There are many other possibilities for them too, such as forming electronic **transistors**, or as parts of computer memory chips.

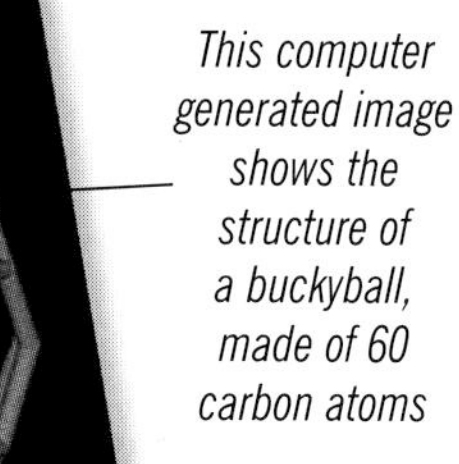

This computer generated image shows the structure of a buckyball, made of 60 carbon atoms

CAN ANYTHING BE HARDER THAN DIAMOND?

In 2005, a research team based in Germany made a new material more than ten percent harder than diamond, the hardest mineral found in nature.

The material is called aggregated carbon nanorods (ACNR). ACNR takes ordinary buckyballs a stage further, creating a new material that is even harder, stronger, and tougher. It is made by heating buckyballs to nearly 5400°F (3000°C), at the same time squeezing them with a massive pressure of more than 1,430 tons (1,300 tonnes).

The diamond is the hardest mineral found in nature

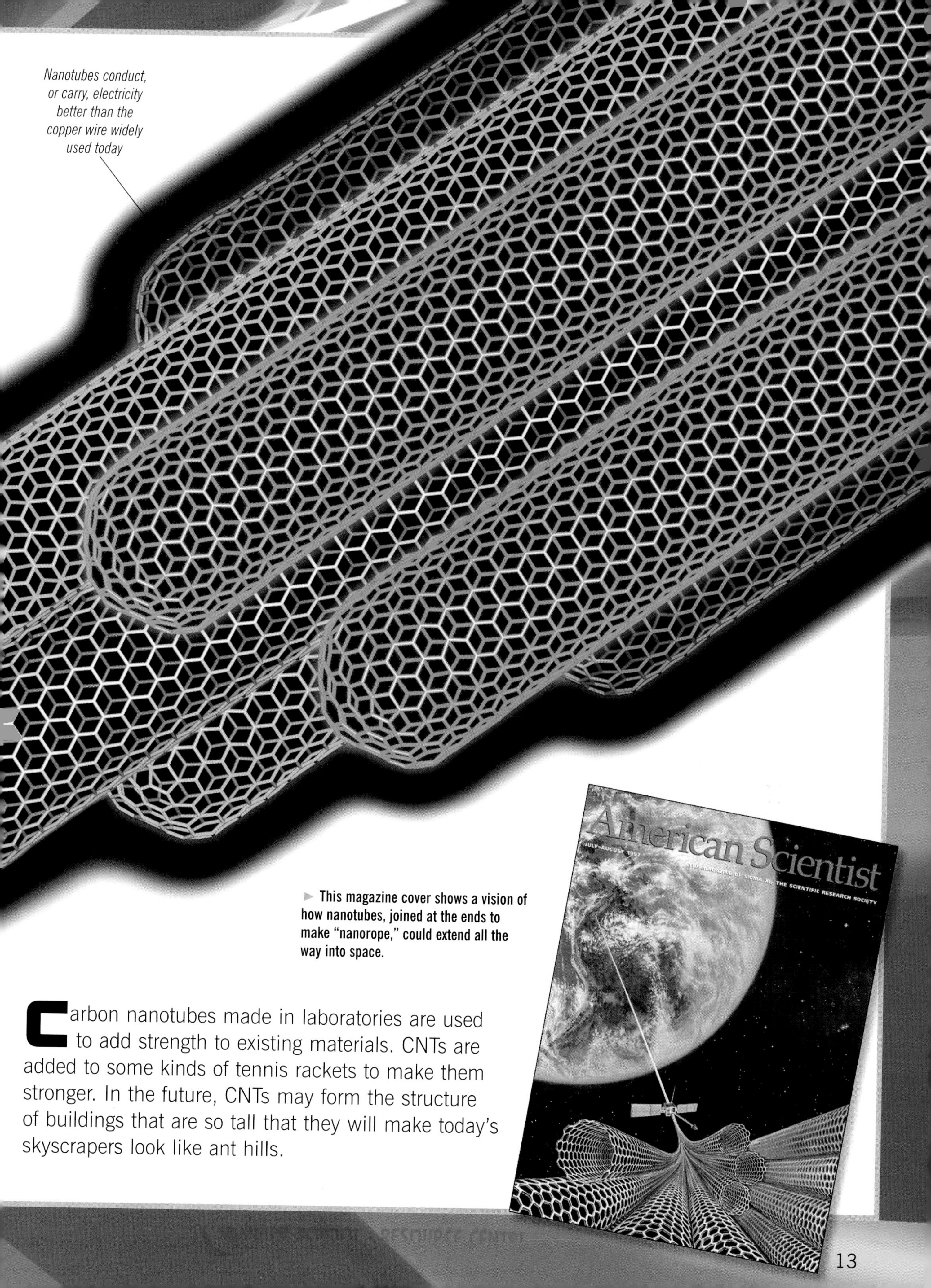

Nanotubes conduct, or carry, electricity better than the copper wire widely used today
American Scientist
JULY–AUGUST 1997
THE MAGAZINE OF SIGMA XI, THE SCIENTIFIC RESEARCH SOCIETY

This magazine cover shows a vision of how nanotubes, joined at the ends to make "nanorope," could extend all the way into space.

Carbon nanotubes made in laboratories are used to add strength to existing materials. CNTs are added to some kinds of tennis rackets to make them stronger. In the future, CNTs may form the structure of buildings that are so tall that they will make today's skyscrapers look like ant hills.

FIRST WAVE OF NANO

Nanotechnology will play an exciting role in the future. For now, nanotechnology is available in stores, in products that anyone can buy.

▲ This lotion has nanoparticles that help it melt into the skin, providing moisturizer with sun protection .

Makers of products that use nanotechnology say that there are benefits to using their products. In the cosmetics industry, nanotechnology is already widely used. Scientists have found that **zinc oxide** paste turns into a silky-smooth clear cream when it is broken down into nanoparticles. These nanoparticles are now used to make sunscreens, and skin moisturizers that have sun protection.

The fashion industry has also developed nanotechnologies, in stain and wrinkle proof clothing. Billions of nanoparticles are added to cotton fabric by dipping it in a special chemical. The particles coat each cotton thread without changing its look or feel. This coating lets liquid roll off the material, without leaving marks, and allows the cotton to resist wrinkling.

Here are some products that currently use nanoparticles.
1 Nanoparticles in skiwax give a super slippery finish to skis.
2 Sun-spray with nanoparticles does not irritate the skin.
3 Nano-core tennis balls seal in air, making them last longer.
4 A golf ball made with nano metal flies in a very straight path.
5 Nano-Tex shirts resist stains and wrinkles.
6 Ski clothes use Nano-Tex to keep the wearer dry and warm.
7, 8 A nanofilm coating protects glass from dirt and scratches.
9 This camera has a screen that uses nanotechnology. It is brighter than earlier displays.
10 Some tennis rackets have nanotube frames for strength.

1

2

◄ Carbon nanotubes (CNT) are used to make parts for mountain bikes. CNT handlebars are about 25 percent stronger and lighter than steel bars.

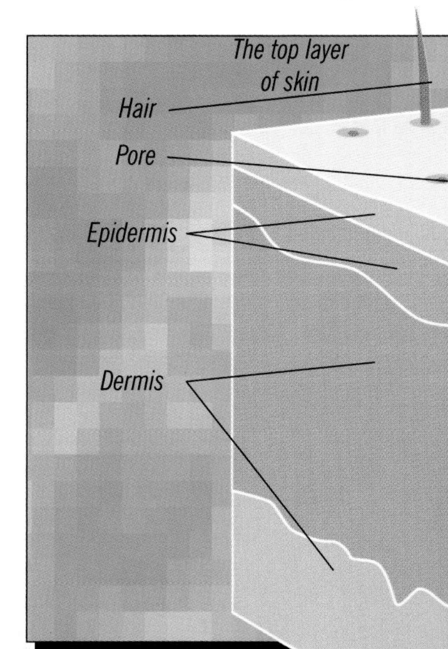

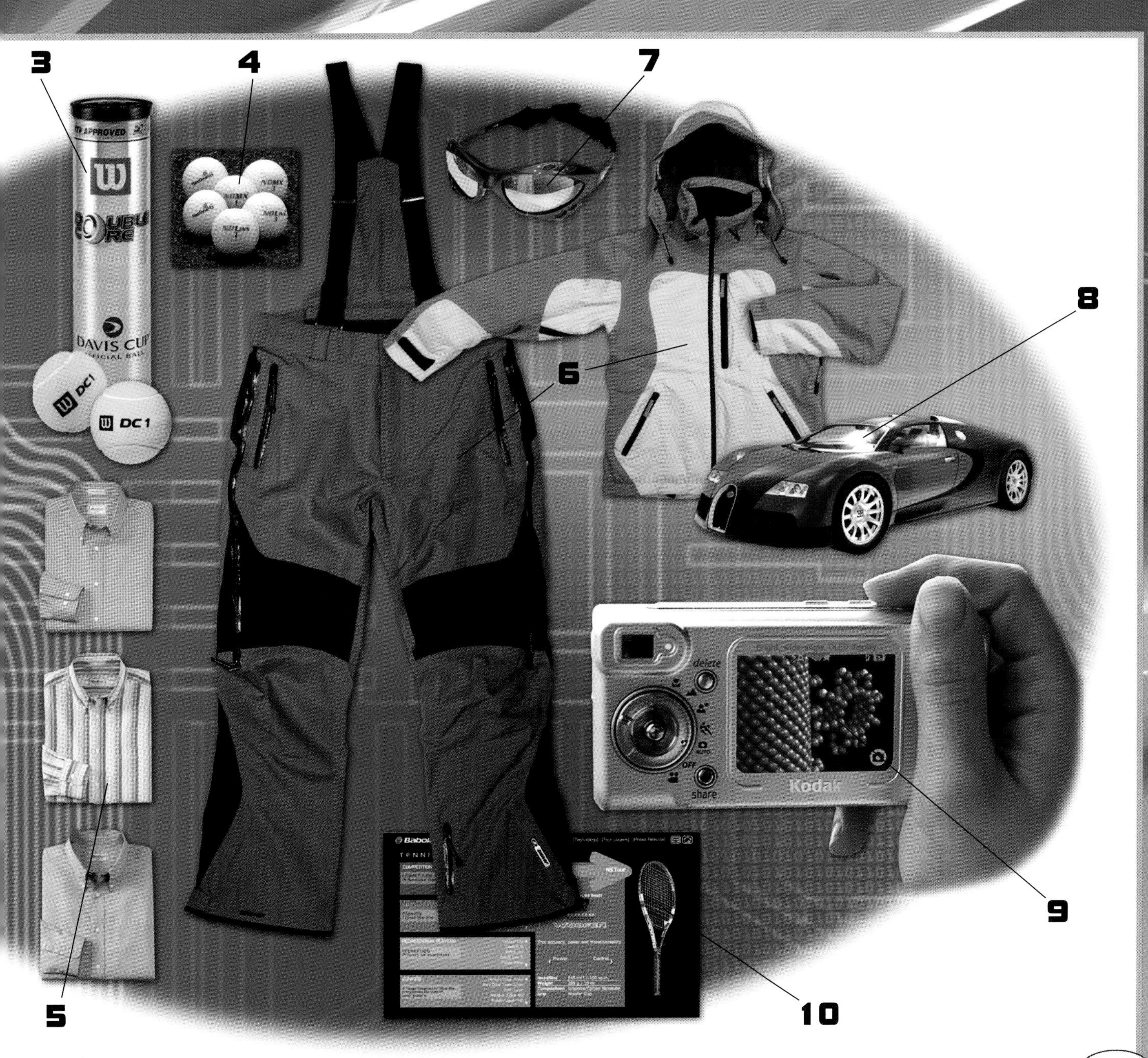

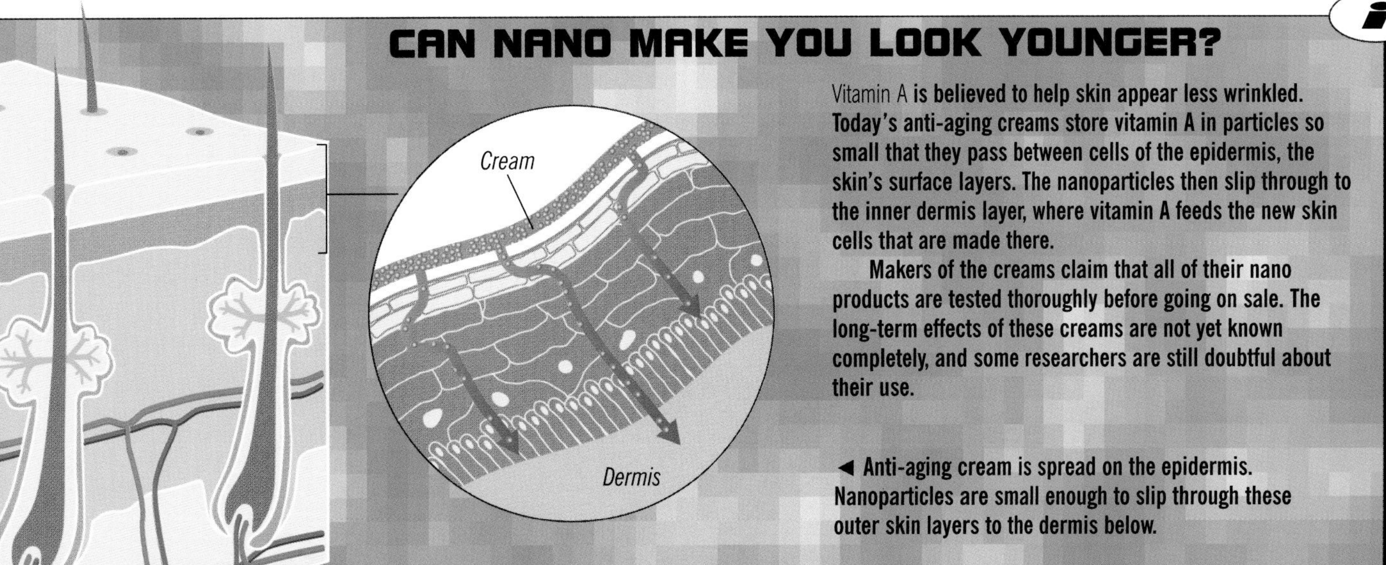

CAN NANO MAKE YOU LOOK YOUNGER?

Vitamin A is believed to help skin appear less wrinkled. Today's anti-aging creams store vitamin A in particles so small that they pass between cells of the epidermis, the skin's surface layers. The nanoparticles then slip through to the inner dermis layer, where vitamin A feeds the new skin cells that are made there.

Makers of the creams claim that all of their nano products are tested thoroughly before going on sale. The long-term effects of these creams are not yet known completely, and some researchers are still doubtful about their use.

◄ Anti-aging cream is spread on the epidermis. Nanoparticles are small enough to slip through these outer skin layers to the dermis below.

NANO DOCTOR

Nanotechnology is already being used in the medical field. The future goals of nanotechnology in medicine include better disease detection, diagnosis, and treatment.

▲ In the future, nanotechnology will help create better replacement organs and body tissue.

Human bodies are filled wilth small structures called cells that help them function. Billions of tiny cells make up a body, and each is made of living nano-sized materials, such as **proteins** and acids. Nanotechnology allows for particles and tools smaller than the cells to help treat patients.

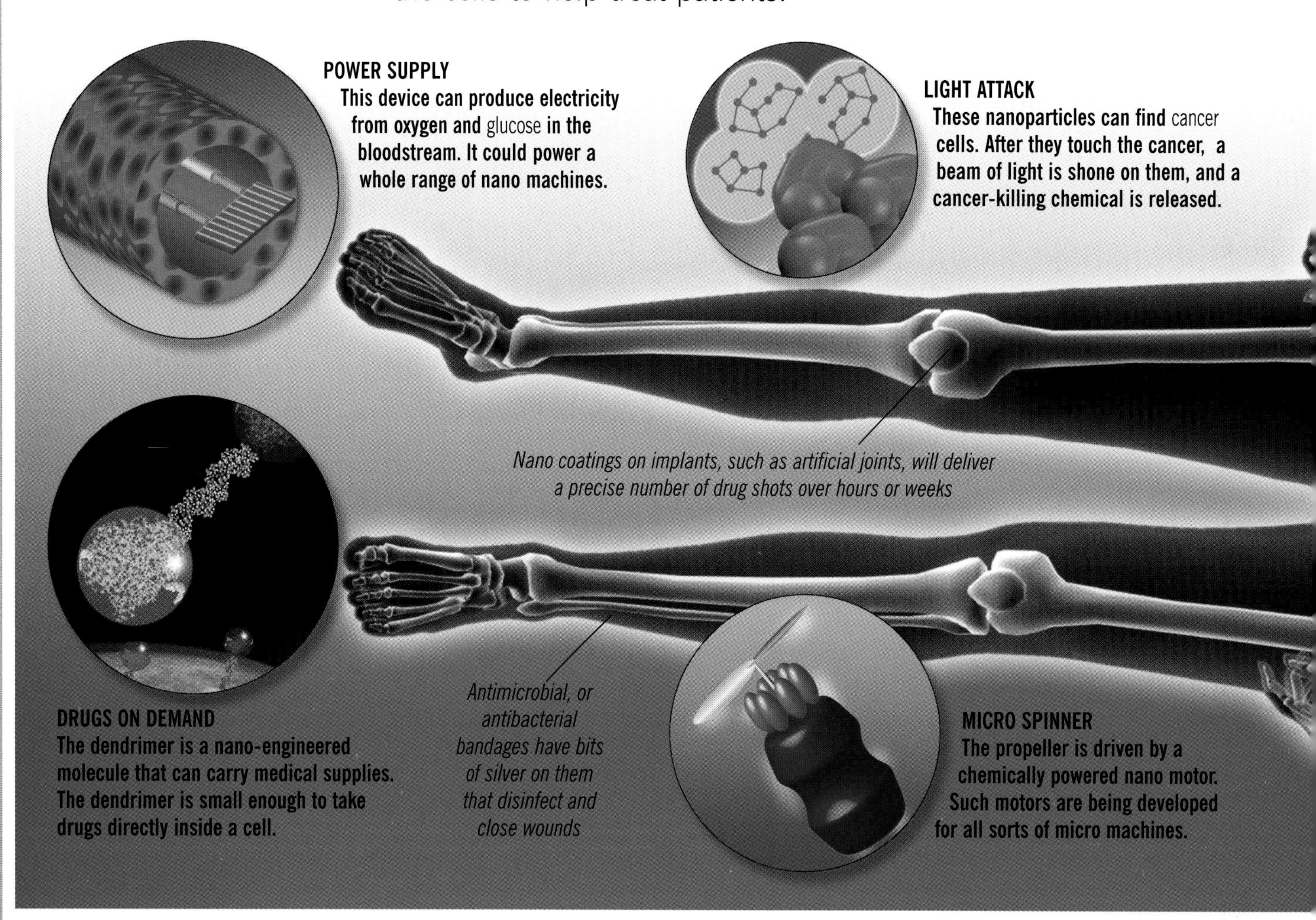

POWER SUPPLY
This device can produce electricity from oxygen and glucose in the bloodstream. It could power a whole range of nano machines.
LIGHT ATTACK
These nanoparticles can find cancer cells. After they touch the cancer, a beam of light is shone on them, and a cancer-killing chemical is released.
Nano coatings on implants, such as artificial joints, will deliver a precise number of drug shots over hours or weeks
DRUGS ON DEMAND
The dendrimer is a nano-engineered molecule that can carry medical supplies. The dendrimer is small enough to take drugs directly inside a cell.
Antimicrobial, or antibacterial bandages have bits of silver on them that disinfect and close wounds
MICRO SPINNER
The propeller is driven by a chemically powered nano motor. Such motors are being developed for all sorts of micro machines.

▲ The nanotechnology research ideas shown here are mostly at the experimental stage. When perfected, they could save many lives.

For something as small as a cell, even the finest scalpel is a huge tool, likely to tear and injure. Today's surgery works largely because cells can usually repair themselves, get rid of dead matter, and then recover by making new tissue.

WHAT WOULD NANO TREATMENT BE LIKE?

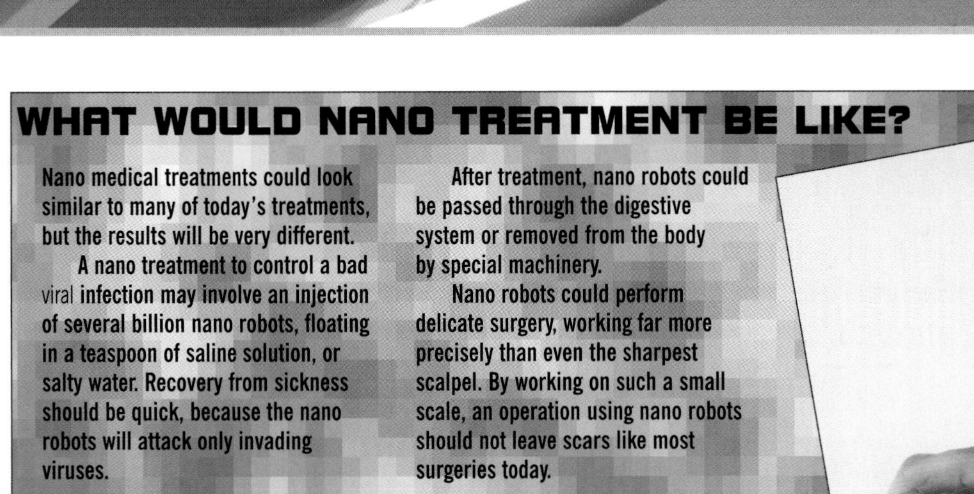

Nano medical treatments could look similar to many of today's treatments, but the results will be very different.

A nano treatment to control a bad viral infection may involve an injection of several billion nano robots, floating in a teaspoon of saline solution, or salty water. Recovery from sickness should be quick, because the nano robots will attack only invading viruses.

After treatment, nano robots could be passed through the digestive system or removed from the body by special machinery.

Nano robots could perform delicate surgery, working far more precisely than even the sharpest scalpel. By working on such a small scale, an operation using nano robots should not leave scars like most surgeries today.

A nano injection will probably look similar to any injection today.

"Flex-Power" nanoparticle cream is now on sale to ease pains in joints and muscles

CELL GRIPPER
This odd-looking machine is a tiny gripper that can pick up objects as small as a single cell.

Future nanosurgery could rearrange the atoms of your ears, nose, or other features

DRUG CAPSULE
This capsule is designed to help people suffering from diabetes. It can check on blood glucose levels and release insulin when needed.

CHEMICAL SUPPLY
This tiny cylinder can detect when living tissue needs more of some vital chemical. It can then supply the needed drugs.

SMOOTH JOINTS
A super-smooth nano coating helps people with hip and other joint replacements. The long-lasting coating lets the artificial joint fit better.

Engineers using nanotechnology will create new instruments for doctors that will let them examine the body in greater detail. Sensors smaller than a cell will let doctors see precisely how the body works. These tiny tools will help make repairs on the body, even on an extremely small scale.

MEDICAL WONDERS

Nanotechnology is developing quickly. Advanced medical micro-electrical-mechanical systems (MEMS) are already being developed in research laboratories.

▲ Respirocytes, a design for a future nanobot, are shown here in blue. They are very strong because of their diamond construction, and can withstand very high pressures.

Researchers are also working on nanobots for use in medicine. Nanobots are a combination of nano machine, robot, and computer. Many different kinds of nanobots will need to be built, as there will be many different jobs to do. A nanobot built to float in the bloodstream could be about 500 to 3,000 nanometers long. Nanobots designed to go through tissues or the digestive system could be around 100 times bigger.

One nanobot, the respirocyte, has already been designed, but has not yet been built. The respirocyte is an artificial red blood cell, designed to carry oxygen in the body. The respirocyte is a hollow, diamond-shaped sphere that is smaller than a blood cell. It could store hundreds of times more oxygen than a real red blood cell. The respirocyte's designer, Robert A. Freitas Jr., believes that an injection of just 0.52 pints (250 milliliters) of respirocytes could have amazing results for patients suffering from respiratory, or breathing, problems. In theory, a person who has had an injection could stay underwater for an hour without needing to breathe!

WHAT IS IT LIKE INSIDE THE BODY?

The microscopic world is no newcomer to science fiction. There have been several movies that have shown what it might be like to travel inside a body.

The first was *Fantastic Voyage*, a 1966 film in which a submarine was shrunk to nano-size, and then injected into the body of a dying man. In the sub was a medical team with a mission to save the man's life. The movie showed scuba-diving surgeons battling white blood cells. The job of white blood cells is to fight infections, or invading organisms. In the end, the heroes hurtled through heart valves and renewed the air supply in the lungs!

In the 1987 comedy *Inner Space*, a miniaturized test pilot was injected into the wrong person by mistake.

In 1989, *Honey, I Shrunk the Kids* did not go inside the human body, but looked at the perils of miniaturization. In the movie, a scientist shrinks his children by mistake. Then they get thrown out with the household trash. Fortunately, the kids survive!

▲ This is how experts think a medical nanobot could look. A nanobot uses an injector (1) to attack a cancerous tumor (2) with a drug that will stop its growth. Gripper arms (3) keep the nanobot in the correct position. The drug is stored in the body (4) of the nanobot.

Medical nanobots of the future might be made of carbon in the form of fullerenes, mainly because of their strength. Other lightweight materials and elements could be used, including hydrogen and oxygen. Silicon could also be used, especially for nano-sized gears and other tiny parts.

MICRO MOVERS

Nanotechnology has already made some changes in transportation, such as the sensors that are used for car airbags. Soon almost every vehicle on land, sea, or in the air, will be using nano systems.

◄ ▼ The arrow points to the panels made using nanomaterials in the Hummer pickup truck. The panels are lightweight and have added strength.

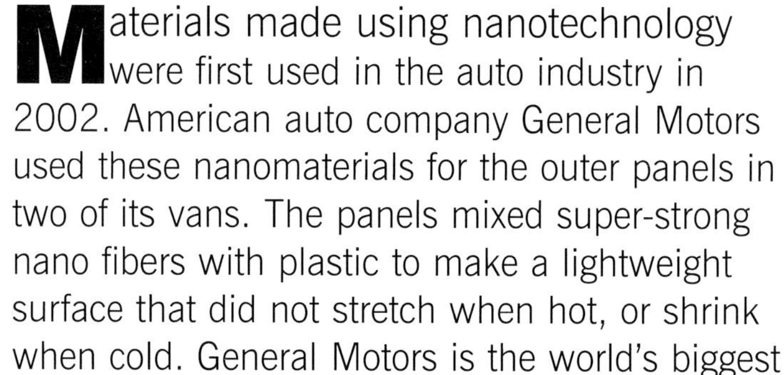

Materials made using nanotechnology were first used in the auto industry in 2002. American auto company General Motors used these nanomaterials for the outer panels in two of its vans. The panels mixed super-strong nano fibers with plastic to make a lightweight surface that did not stretch when hot, or shrink when cold. General Motors is the world's biggest user of nanomaterials, using more than 330 tons (300 tonnes) per year.

A special polish made with nano particles keeps vehicles looking as if they were newly painted. The nanoparticles form a barrier against road grime, dirt, salts, and the fading power of the Sun.

Painting a ship's hull with a super slippery nano coating allows the vessel to slice through the water a little faster.

The coating also means less expensive maintenance. Barnacles and other sea animals cannot stick to the slick surface as well as to a hull coated with normal paint.

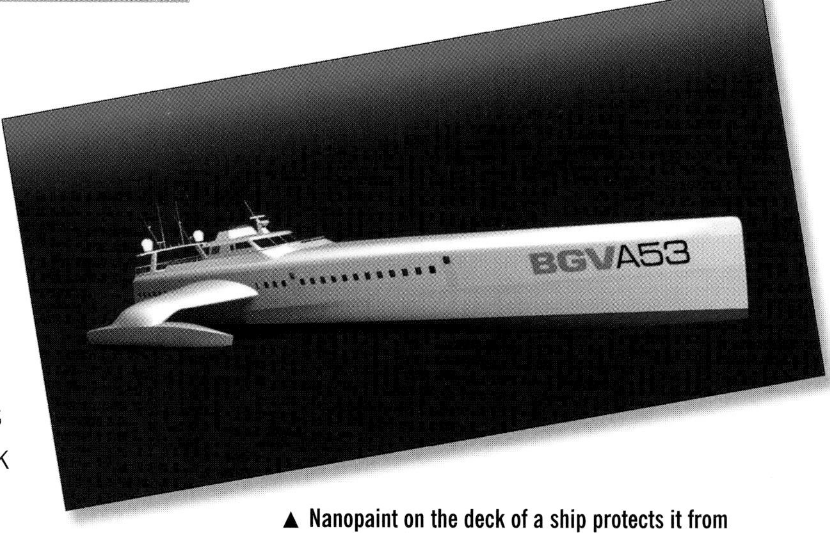

▲ Nanopaint on the deck of a ship protects it from the harsh effects of sunshine and salty sea spray.

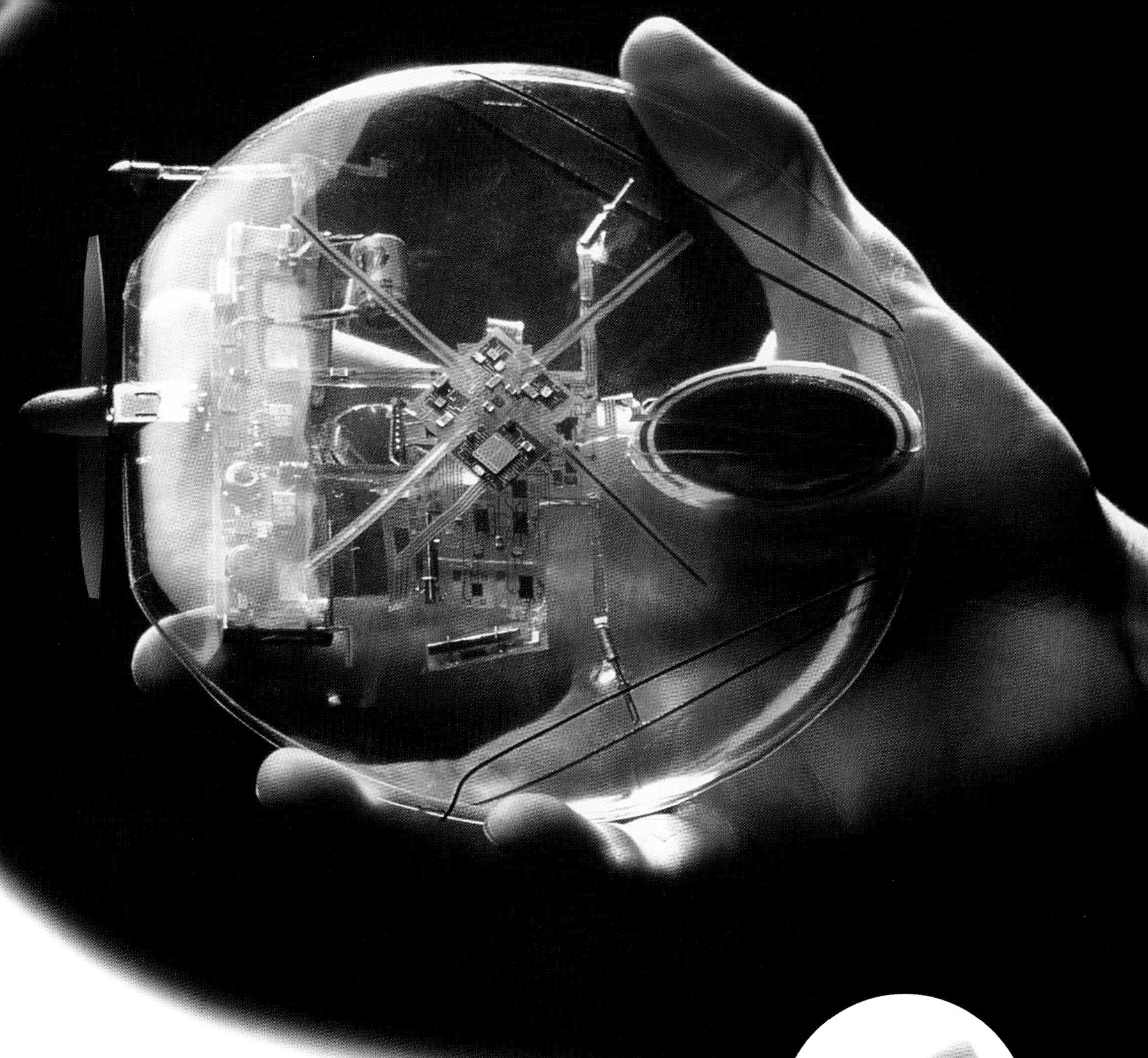

▲ The Black Widow micro air vehicle (MAV) is just six inches (15 centimeters) wide and weighs barely two ounces (60 grams). It flies at 44 miles per hour (70 kilometers per hour) and can stay in the air for an hour at a time. Micro machines like this can use small cameras to act as spies.

Aircraft designers are trying to find ways to build lighter planes that carry more weight and use less fuel. Nanomaterials allow attack and spy planes to be built smaller. The smaller they are, the harder it is for an enemy to spot them.

IS A FLY-SPY POSSIBLE?

A robofly shown at actual size

A new field of study, called biomimetics, aims to build robots that copy the actions of living creatures, such as diving like fish, crawling like ants, or zipping around like flies.

The robofly was the result of three years of hard work by an American research team. It measures just one inch (2.5 centimeters) wide, and has wings similar to a fly.

The robofly has two tiny motors that make its foil wings flap and rotate, much like real flies. So far, the robofly can only move forward, but the designers are aiming for a future model that will be able to move in all directions.

EXPLORER SWARMS

In the past, exploration vehicles were big and bulky. Nanotechnology will allow them to be made smaller, smarter, and cheaper.

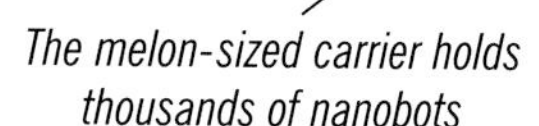

The melon-sized carrier holds thousands of nanobots

▶ A probe releases a swarm of nanobots. They are light enough to be carried through the air for long distances.

The nanobots have a common design, but they carry different sensors, including pinhead-sized cameras

▲ Two Rovers landed on Mars in 2003. They drove slowly across the rusty deserts under direction from controllers on Earth.

The size and weight of cargo contribute to the cost of a space mission. The bigger the cargo, the bigger the rocket needed to blast it into space. Two Mars Rovers, each the size of a microwave oven, have explored parts of Mars. They were miracles of miniaturization when built, but future nano machines will make them look like lumbering elephants.

A ballbot is an idea for a future exploration machine. Swarms of the marble-sized robots could be sent to other planets. The plan is to build different types of ballbots, such as photographers and soil-samplers. Their small size and ability to move well would allow them to explore caves, deep valleys, and other remote spots.

▲ Ballbots would hop by sending a surge of power to a "foot," leaping three feet (one meter) at a time.

Nanobots are designed to dissolve safely after the mission, to avoid polluting the environment

To sample the gases of another planet, or a volcano on Earth, a cloud of feather-light nanobots could be released. Shaped like the seeds of a sycamore tree, they would drift in the breeze, reporting to researchers using tiny **radio transmitters**.

INSECTS ON THE LOOSE?

Future robot swarms may be based partly on the work of artificial intelligence researchers during the 1990s. "Elma," a robotic insect, was designed to learn about its environment using a simple computer program.

The team that developed Elma built it to operate much like a real insect. Despite being simple creatures, many insects develop astonishing behavior patterns. For example, termites build huge hills that would seem to be beyond the capabilities of their small brains. Elma teaches itself to walk in just 12 days.

Elma was an early walking robot. It had special sensors in its "head" that used sound waves to check how far an object was from it.

NANO DOOM?

▲ The idea of nanotechnology taking over the world appeals to science fiction authors. In his novel *Prey*, Michael Crichton writes of a terrifying world in which a swarm of microscopic killer machines goes out of control.

Some people are worried about the effects of nanotechnology on the environment. Are they right to be concerned?

In 1986, engineer Eric Drexler warned that nano machines could get out of control, possibly affecting life on Earth. Drexler suggested that they might wipe out living matter, leaving behind a dusty waste called gray goo. Most scientists now believe that gray goo is unlikely because MEMS and nanobots are not advanced enough.

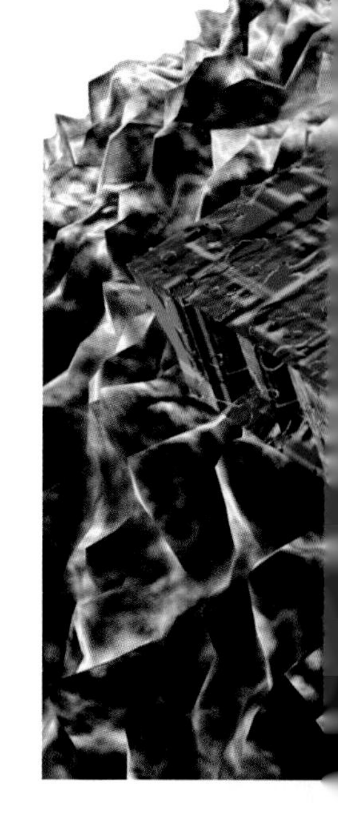

◄ Nanoparticles are all around us. They are in vehicle exhausts, smoke from fires, and chimney fumes. They are even given off when frying food in a hot pan.

Nanoparticles exist naturally and can also be created in laboratories. Smoke, vehicle exhaust, and other pollutants are examples of nanoparticles that we breathe in every day.

▲ One fear is that nano machines could have the ability to reproduce themselves repeatedly. Some people think that vast armies of nanobots could become a deadly danger to anything alive. Scientists say this is unlikely to happen.

ARE NANOBOTS JUST A SCARE?

Richard Smalley was on the team of researchers that discovered fullerenes. According to Smalley, nano-sized robots are unlikely to be much of a threat. Even if nanobots could assemble a billion atoms per second, then it would still take a long time to build anything. At this speed, says Smalley, a mouse-sized object would take many years to finish.

There are concerns that nanoparticles made in laboratories could escape into the environment. There, they might come into contact with plants and animals that have never been exposed to nanoparticles. The effect that nanoparticles could have on these plants and animals is unknown. They may cause damage to animal lungs making it difficult for them to breathe. They could also damage or kill plants.

COMING SOON?

Soon, we will live in a world where nanotechnology devices are all around us, and probably inside us, too. Will it be a better world? Only time will tell.

Nanotechnology may solve the serious world problem of a shortage of clean drinking water. At present, 1.3 billion people lack a regular supply of clean water, and the rising world population means that water needs may double by 2030. The situation is so bad that some people think there may be wars in the future for control over water supplies, such as rivers or lakes.

▲ Nanotechnology filters already help provide clean water in Asia. They reduce pollution, by sifting out poisons from the water supply. Hopefully, tragedies like the poisoning of these birds will become rare in the future.

Water filters made from nano materials help supply clean water. These are the first filters that can block harmful bacteria and viruses effectively. In Bangladesh, nanofilters sift out deadly arsenic, a poison that occurs naturally in the water supply.

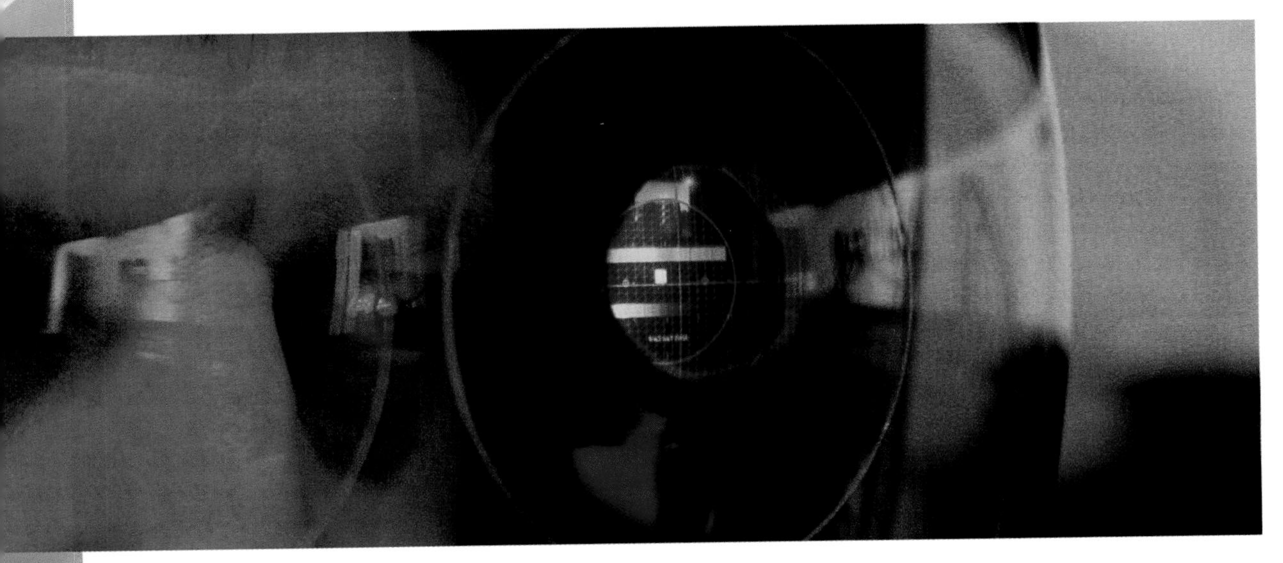

◄ Air forces will soon use nanotechnology to link pilots with their combat computers. This should improve accuracy, by making weapons easier to control.

Looking further into the future, a team presently working on nanobots predicts that nanotechnology will boost our mental abilities. The team believes that just one nanocomputer, the size of a human body cell, could store as much information as the world's biggest library. If ever perfected, the nanocomputer could be inserted into a human, making that person super-smart!

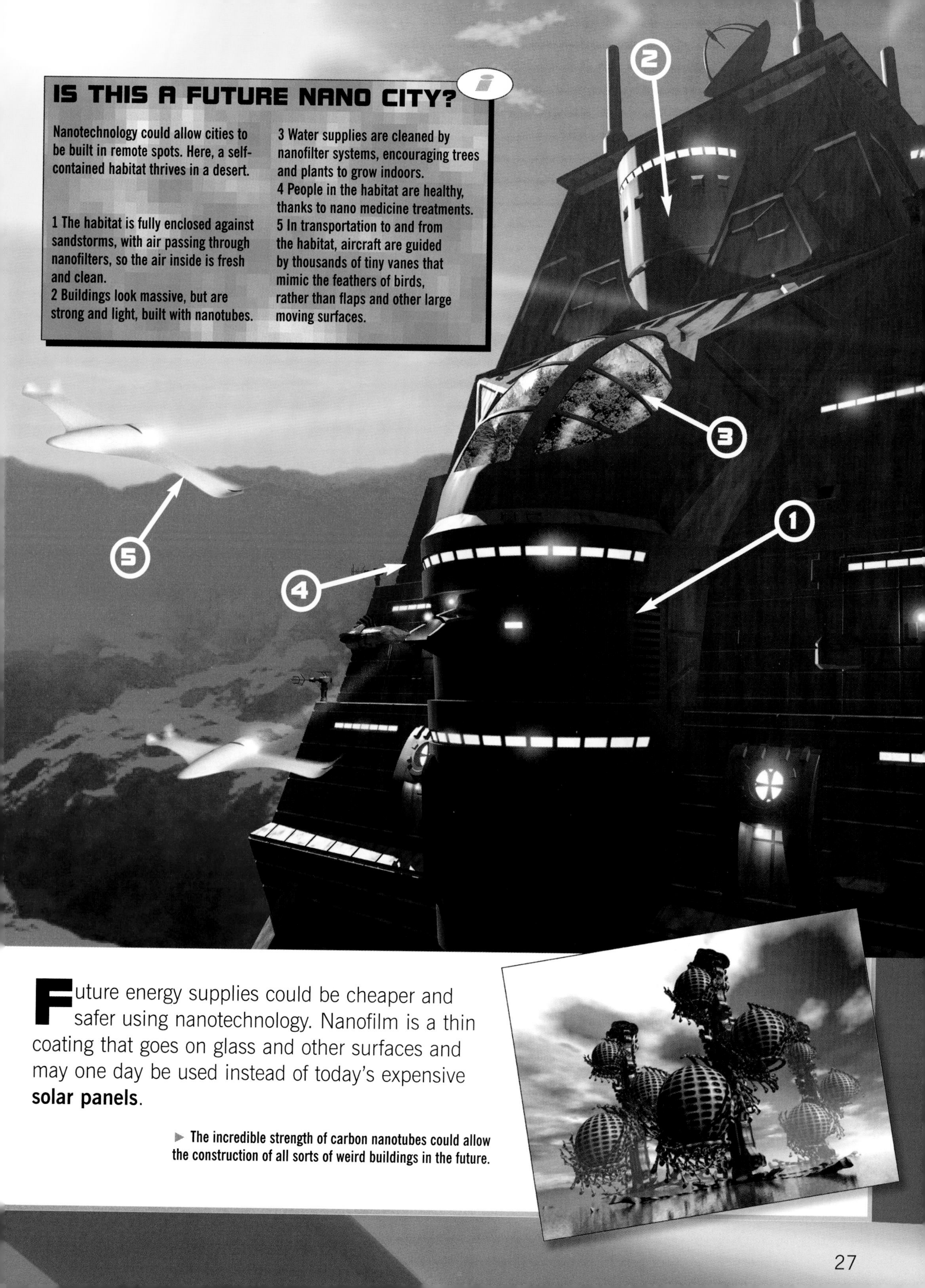

IS THIS A FUTURE NANO CITY?

Nanotechnology could allow cities to be built in remote spots. Here, a self-contained habitat thrives in a desert.

1 The habitat is fully enclosed against sandstorms, with air passing through nanofilters, so the air inside is fresh and clean.
2 Buildings look massive, but are strong and light, built with nanotubes.
3 Water supplies are cleaned by nanofilter systems, encouraging trees and plants to grow indoors.
4 People in the habitat are healthy, thanks to nano medicine treatments.
5 In transportation to and from the habitat, aircraft are guided by thousands of tiny vanes that mimic the feathers of birds, rather than flaps and other large moving surfaces.

Future energy supplies could be cheaper and safer using nanotechnology. Nanofilm is a thin coating that goes on glass and other surfaces and may one day be used instead of today's expensive **solar panels**.

▶ **The incredible strength of carbon nanotubes could allow the construction of all sorts of weird buildings in the future.**

MICRO MACHINES TIMELINE

Here you can track the progress of nanotechnology and micro machines from early ideas to present reality.

▲ Richard Feynman was one of four American scientists who were featured on a special stamp issued in 2005.

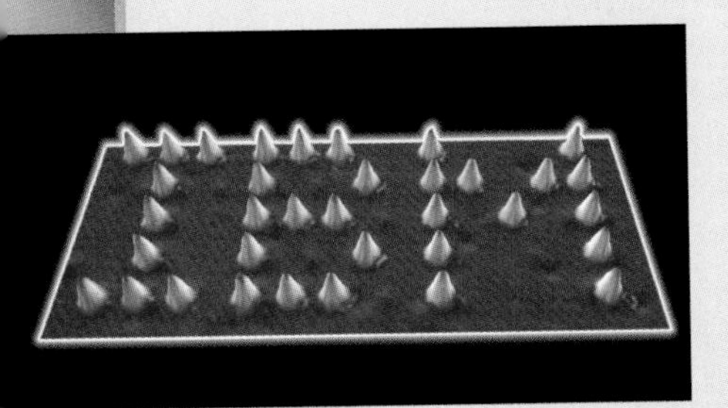

▲ The letters IBM were made from 35 atoms of the element xenon, in 1989. It is shown here magnified many times.

1955 The first images of individual atoms are made, using a recently developed instrument called the field ion microscope. Scientists note that atoms are not solid, and are made of electric charges.

1959 American physicist Richard Feynman (1918-1988) gives a talk called *There's Plenty of Room at the Bottom*. In it, he suggests putting the contents of a set of encyclopedias on a pinhead. This talk is believed to be the real start of nanotechnology.

1960 An electric motor is made that is less than 0.156 inches (four millimeters) in size. It wins a $1,000 prize offered by Richard Feynman in his 1959 talk.

1969 Early micro-electrical-mechanical systems (MEMS) are used in transistors made by the American Westinghouse company. Today MEMS are everywhere, from airbag sensors to ink-jet printers.

1974 Tokyo Science University professor Norio Taniguchi coins the term "nanotechnology." He uses it to describe manufacturing parts to an accuracy of 1,000 nanometers or less.

1981 The scanning tunnelling microscope (STM) is invented by a research team at IBM. The STM can show images of molecules and atoms.

1985 Another prize offered by Richard Feynman during his famous 1959 talk is claimed when a book page is reproduced 25,000 times smaller than life-size.

1985 Fullerenes, a new form of carbon molecule, also known as "buckyballs," are discovered by Richard Smalley, Sir Harold Kroto, and Robert Curl, working at Rice University in the United States.

1986 K. Eric Drexler (1955-) writes the *Engines of Creation*, a book which maps out the new world of nanotechnology and how it might develop. Drexler also wrote *Molecular Engineering* in 1981 and *Unbounding the Future* in 1991.

1986 The atomic force microscope (AFM) is perfected. Not only can it display images of individual atoms, it can also be used to place and move them. The AFM has since become one of the most important tools used in nanotechnology.

1989 Researchers with the IBM computer company use an atomic force microscope to arrange 35 atoms of xenon gas, spelling out IBM's name. The tiny object becomes the world's smallest company logo.

1991 The carbon nanotube, an elongated fullerene, is discovered by Sumio Iijima (1939-) of Japan.

1997 Zyvex, the world's first company dedicated to nanotechnology, is formed. Today, Zyvex has many nano products for sale, such as carbon nanotubes, which add strength to sporting goods.

1997 A team at Cornell University in the United States makes the world's first nanoguitar, an instrument too small to be seen by the naked eye. There are six "strings," each of which is only about 50 nanometers wide.

1997 A design for a nano robotic "hand" is developed, which could be made from only 3,000 atoms.

2000 Early nano products go on sale to the public. They include sunscreens and hair dyes.

2002 "Nano-care" clothing, which is coated in nanoparticles to make stain-repellent fabrics, is developed. The nanoparticles that coat the fabric are just ten to 100 nanometers long.

2004 The dangers of nanotechnology are highlighted by a study that shows fish becoming brain damaged after two days of exposure to carbon fullerenes in their water. Fullerenes killed water fleas in other tests. The researchers think that, although living systems can get rid of existing nanoparticles, new nanoparticles may present problems.

2005 A study of possible environmental dangers of nanotechnology by a Swiss insurance company takes place. The results show dangers that are similar or less dangerous than other industrial processes, such as oil refining.

2005 First big use of the word "nano" in a consumer product. Apple computer company puts its iPod Nano music player on sale. The Nano is not really nano-sized, but it is still extremely small, measuring just 3.5 inches (nine centimeters) tall.

2005 Production of nano materials throughout the world continues to rise. Altogether, their value is more than $23 billion. Experts believe this will double in the next five years.

2006 Further progress in making MEM systems comes with the smallest hole ever drilled. A science team based in Wales makes perfect holes as small as 22,000 nanometers wide, roughly one-quarter the width of a typical human hair.

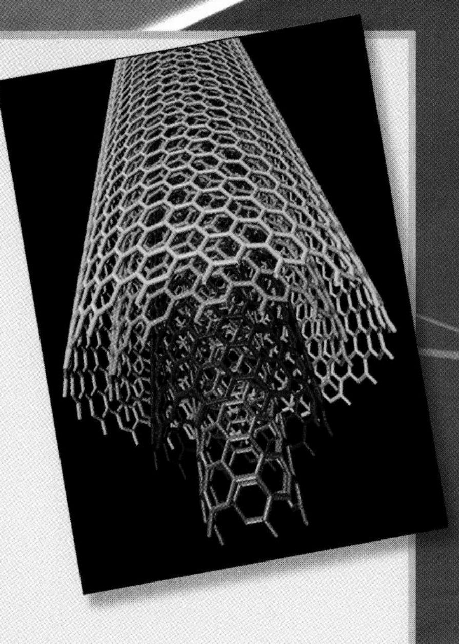

▲ **Carbon nanotubes can be constructed in several layers, as shown here.**

▲ **Cornell University's 1997 silicon nanoguitar was about the size of a red blood cell. In theory, the strings could be plucked by an atomic force microscope, but the sound would be too high-pitched for a human to hear.**

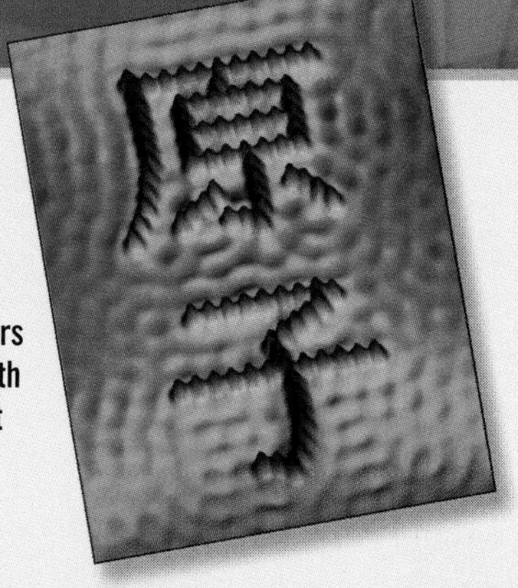

▶ The Japanese letters for "atom," written with atoms of iron, laid out on a copper sheet.

GLOSSARY

An explanation of some technical words and concepts used in this book.

Amoeba A single-celled animal that moves around by extending finger-like extensions called pseudopods.

Artery Any of the small tubes that carry blood away from the heart.

Bacteria Single-celled organisms that reproduce by dividing. Many types cause illnesses, but other bacteria, such as those that change milk into yogurt, are beneficial.

Biological systems Living organisms and how they function.

Cancer A deadly disease caused by abnormal cell division, or growth.

Carbon A non-metallic element found in two main forms, diamond and graphite. All living things contain carbon molecules.

Carbon dioxide A colorless gas in the air. The carbon dioxide molecule is CO^2, or one carbon and two oxygen atoms.

Cell The basic unit of a living thing. Most plants and animals are made of many cells of different shapes and sizes, carrying out various functions.

Digestive system A series of organs in humans and other animals that help break down food and release waste.

Dust mite A tiny animal from the same family as spiders and scorpions. Dust mites eat flakes of dead human skin.

Engineering The planning, designing, and building of structures and machines.

Glucose A simple sugar that is used by the body to create energy.

Insulin A substance produced by the human body that helps regulate the amount of glucose, or sugar, in the blood.

Mass The measure of how much matter an object contains.

MEMS Micro-electrical-mechanical system, a term used for extremely small machines and mechanisms.

Nanoparticle A microscopic particle smaller than 100 nanometers. It may be manufactured, such as a fullerene sphere, or natural, such as particles of smoke.

Physicist A scientist who studies matter and energy.

Protein A complex molecule that makes up many structures in the body, such as hair and muscle.

Radio transmitter A device used to send signals by a type of energy called radio waves.

Red blood cell The cells in the body that take oxygen to tissues and organs.

Rudder A device on a boat or plane used for steering.

Scale The size of an object in relation to other objects.

Sensor A general term for mechanical devices that perform similar functions as the human senses of touch, taste, sight, sound, or smell.

Silicon The second most abundant chemical element found in the earth. Silicon is often used to make transistors.

Solar panel A flat piece of material that converts the energy in sunlight to electricity. This process is called the photovoltaic (PV) effect.

Sound waves Energy that carries sound through the air.

Surface tension An effect within a liquid's surface that causes it to behave like an elastic sheet. Surface tension allows insects to walk on water.

Teflon A nylon material often used to give equipment, such as cookware and medical instruments, a non-stick surface. The name is short for polytetrafluoroethylene.

Transistor An electronic part used in computers, radios, and similar devices. It can act as a switch to amplify sound, and for many other functions. In the future, nanoelectronic systems may use fullerene spheres to replace the much larger transistors used today.

Van der Waals force An attraction that draws molecules together when they are close. The force is named after the Dutch scientist, Johannes Diderik van der Waals (1837-1923).

Viral, Virus A harmful molecule that attacks a body's cells and makes a person sick.

Vitamin A Also known by its scientific name, retinol, vitamin A is a yellow chemical found in egg yolks, fish liver oils, and many vegetables. It is essential for cell growth in our bodies and also helps our vision.

Zinc oxide A solid white substance used as a pigment in some paints, as well as sun protection in some ointments and sunscreen lotions.

▲ A MEM set of gears, shown here balanced on a fingernail.

GOING FURTHER

The Internet is a good way to keep up with the latest in nanotechnology and micro machines. Here are some sites to get you started.

http://www.nasa.gov/vision/earth/technologies/27jul_nanotech.html
This page is part of the NASA site, and gives you an idea of what the future holds for nanotechnology in space.

http://science.nasa.gov/headlines/y2002/15jan_nano.htm
Another NASA page, with fascinating information on nano developments.

http://www.zyvex.com/nanotech/feynman.html
This page contains the text of Richard Feynman's 1959 talk.

http://science.howstuffworks.com
A general science site that has information for everyone who is interested in the "how" and "why" of things.

For more nanotechnology information, remember to use your school or local library. Nanotechnology is a hot topic that is constantly changing, so the popular science magazines listed below are a good way of keeping up to date with the latest news.

Scientific American
Probably the best-known of all science magazines, it is packed with amazing articles, and great color illustrations and photos.

New Scientist
A big-selling weekly magazine that is readily available. It has excellent news pages and science articles.

INDEX

Acknowledgements
We wish to thank all those individuals and organizations that have helped to create this publication. Images were supplied by: American Scientist magazine, BAe Systems Plc, Tony Bostrom, Andrew Caballero-Reynolds, Cornell University (nanoguitars), Tobias Dahlen, DaimlerChrysler Corp, Christian Darkin, Fotolia, Sheldon Gardner, Jose Gil, General Motors Corp, Steve Gschmeissner, Victor Habbick visions, Roger Harris, Patrick Hermans, Eric Herve, IBM – Lutz & Eigler, iStockphoto, David Jefferis, Michael Knight, Morgan Mansour, Rory McLeish, Peter Menzel, NASA Space Agency, A. Pali, PSA Peugeot Citroen, Sandia National Laboratories, SUMMiTTM Technologies, www.mems.sandia.gov, Science Photo Library, U.S. Post Office

1 2 3 4 5 6 7 8 9 0 Printed in the U.S.A. 5 4 3 2 1 0 9 8 7 6